ANIMALTALK 101

LEARNING TO SPEAK TO ANIMALS

LAUREN BODE AND JOHN BELL

authorHOUSE™

1663 LIBERTY DRIVE, SUITE 200
BLOOMINGTON, INDIANA 47403
(800) 839-8640
WWW.AUTHORHOUSE.COM

First published by AuthorHouse 12/27/05

ISBN: 1-4208-8536-7 (sc)
ISBN: 1-4259-1178-1 (e)

Printed in the United States of America
Bloomington, Indiana

This book is printed on acid-free paper.

Table of Contents

Introduction

by John Bell

How Lauren's Animal Telepathic Ability Works

Love of animals is a universal impulse,
a common ground on which all of us
may meet. By loving and understanding
animals perhaps we humans shall
come to understand each other.
Louis J. Camuti, D.V.M.

Lauren is a true telepath and has been throughout her whole lifetime. She has no single dramatic incident in her past that "changed her life". She does not claim to have had a "near death experience" which brought "enlightenment". She just has this ability to be aware of thoughts, feelings, images and emotions that originate elsewhere, beyond her physical body, beyond her physical location and even beyond present time.

This sensory anomaly was not previously unknown in her family. Relatives on her father's side have exhibited similar unusual abilities, often involving intuitive healing. Her father, possibly the most influential person in her young life, did nothing to discourage her uniqueness. His "princess" was taught that there were no limits to her abilities and to trust her own judgment. He protected and defended her from anyone who tried to tell her what she could or could not do, and quickly dealt with neighbours and so-called authorities (educational, religious or just intrusive) who had the audacity to criticize or blaspheme her talents. And there were enough of those. Frightened, jealous or superstitious peers and elders had called her many things. Thus supported by her family, Lauren grew as a confident self-reliant entrepreneur.

The imprint of these formative years is evident in her today. Lauren is not one to stand in anyone else's shadow. Her bearing and attitude are assured and even regal, but not in any way condescending or belittling of others. Her assurance in her abilities does not allow room for question or demands for proof. Clients, who use trickery or lies to try to trip her up, find that they have just had their last session with her. Nothing gives her greater joy than the successes of her clients, but she must be respected and given her due. She finds it very irritating when clients consult her in secrecy and then deny any knowledge of her existence around family or religious leaders.

Specific details amaze

The clearest evidence of the extent of Lauren's telepathic ability is in her confident delivery of specific names and details. She regularly astounds her clients by describing clearly things known only to themselves and not mentioned even to their families. I find it amazing that Lauren is able to pass on this information without the "normal" mental screening that might tell her that this is totally improbable and laughably presumptuous.

Once when Lauren was reading by telephone for a new Californian client's first call, she declared: "Frankly, dear, you are overmedicated, and there is something wrong with your right leg. No, wait, you are just adjusting your pantyhose." There was a slight delay in the reading until the client stooped to pick up the phone that she dropped and admitted that she was fighting a cold with remedy that was making her sleepy and her unruly hosiery was now in place so that she could sit down.

Very recently, Lauren spoke to another new client in New York City who had sent a photo of herself prior to the reading (a helpful but not necessary action).

"Your husband has a very unusual nickname and I hesitate to say it because I never heard of a man called that."

"Go ahead. Tell me", laughed the client.

"It's Badger!" chuckled Lauren.

"That's right!" gasped her client.

One minute later in the middle of a question and answer period, Lauren interjected, "Your father who passed away five years ago is with you now and he says he really approves of your new car."

The client confirmed that "Dad" had been dead for five years and it was just like him to be impressed by her vehicle. "We just got a new blue one with blue leather upholstery. Dad's favorite colour."

Lauren performs with the same facility extracting specific names and events from animals. While working with horses, she will pull a name (of a person or another horse) or a specific incident right out of the blue, fearlessly. She operates without the safety net of "an analytical probability filter" on her statements, and perhaps it is best. The owners or riders are jolted by the exact details that "she couldn't possibly know", and then they drop their disbelief, allowing the communication to proceed at a faster pace.

Animals too are sometimes startled
Sometimes the "disbelief" is on the part of the horses. Often they are suspicious of this strange visitor who often is accompanied by a throng of onlookers. Is it a Vet? Are they going to stick something in me? Is it a prospective buyer? Does this mean that I have to move again? Why the crowd? What have I done?

Horses will often be nervous at first and skeptical that "this woman is actually listening to me". After a few moments horses will accept Lauren and start to reveal themselves in their true personalities. It is amazing how different they are. Some are proud and aristocratic, others meek and subdued. Some are overly aware of their athletic ability and tend to deride (pun intended) their human partners as not really a good match for them. Others have such a bond with the owners that they express great concern about the trials and tribulations of their two legged comrades.

We have met an aristocratic stud that resented going back to the same mare. It should just be necessary for only one cover.

A pony of the same age as its young owner with such a loving bond that it would do anything for the girl.

A much sold gelding who said in a macho attitude that he just didn't care anymore. "So what! I can survive anything."

Hypochondriacs, complainers, bullies and victims, mother figures, and usually two or more herd bosses in each stable.

Animals are the real clients
It is such a pleasure to meet these beasts that Lauren occasionally finds it necessary to edit their communications. Certainly the humans are the paying clients, but the communication brings a special empathy with the horses. She feels that she must respect the privilege of sharing the horses' thoughts. Horses will sometimes tell her things that are not to be passed to the owners, or occasionally she will feel that direct presentation of the horses' feelings would not be in the best interest of the animals, so paraphrasing or even hedging on the message is appropriate. This does not happen often, but it can, and Lauren makes no apology. Betraying the trust of an

animal might end the session right there, and might make further work in that stable futile.

The horses react to Lauren differently depending on each individual's situation and temperament. As said earlier, often the horses are nervous at the start of the session. By the time Lauren is through, they can exhibit unusual physical participation in the conference. One might nod vigorously as information is passed, another may nudge Lauren to express urgency, still another might paw the ground in a "let's get on with it" attitude, and there are those who will stand trancelike caught up in the feelings. And they do project feelings. Almost all observers should be able to pick up some of the emotional states of the horses. Certainly any alert owner or trainer should feel with their horses the emotion behind Lauren's transmitted messages.

Horse power
There is a power about horses that attracts so many people. Their physical presence alone, so tall, heavy and muscular, stirs a feeling within that can affect in many ways. With some people, understandably, it is fear. With others, there is a touch of envy. Horse people though usually have another feeling, an attraction that says that in some way "I can be part of this power", "I can manage it, direct it and even become one with it". From ancient times, a person on horseback was lifted above the mundane and into a greater range of power. Very heady stuff! And of course, paying for and caring for such a huge compliant beast leads to an emotional bond. Riders' attitudes can

range from an "I'll show this brute who is boss" through to an affinity that quickly forgives missteps, delivers words and extra caresses of praise and elicits deep concern over the steed's discomfort or injury. Across that wide band of attitudes, any rider worth his saddle opens up to have "a feeling" for the mount. And a type of rapport is quickly agreed upon by both horse and human, often before the butt hits the saddle.

So if all others can develop a sensual feeling for these animals, think of how horses affect an open telepath like Lauren. In most stables, Lauren will get a joyful strength from these powerful but sensitive creatures. But it only takes one or two who have been ill, injured or in conflict to upset her immensely. She may get a headache, a pain in her body corresponding to the animal's discomfort or a feeling of extreme depression appropriate to the horse's state.

One wintry Sunday, Lauren approached her first horse at a large stable, and on touching the animal, staggered back dazed and holding her head. "This animal is sick!" she told the owner. The owner agreed and said that the vet was due to follow up on treatment the next day. The owner only half-apologized for not warning Lauren, as apparently it had been planned as a test of Lauren's perception.

Lauren quickly gave a summary of the horse's symptoms and suggested that something be done for the pain that the horse was feeling, and said that she would come back to this horse after reading for the others. She could not stay with the ailing horse any longer because she was picking up an acute headache and stomach cramps. Later when she returned to that stall, the horse told her that it was afraid that it had been over-medicated. It said that it had received three doses of medication. The owner confirmed that normally they gave the horse only two shots for its ailment, but today, after seeing Lauren's reaction, they gave an extra dose. Lauren completed the reading by giving the owner a list of symptoms that the horse was feeling which was copied down on a pad of paper for the Vet's reference. She then reviewed the health history of the animal and assured it that the owner was now aware of the details and would take necessary corrective measures.

Lauren is not a veterinarian and will not prescribe any specific medications, but she will, as in this case, interview the animal to obtain information about symptoms that otherwise would only be guessed at by the owner or medic. She was quite upset about this incident, because she was not forewarned or prepared for the discomfort that she suffered through the animal. It is not fair or funny to subject anyone to the nauseous ailments of a powerful beast without sufficient warning and time to prepare personal defenses. Of course, Lauren does encounter ailing animals and talking to them is one of her most valued services, but forewarned, she can approach with her defenses up. That is, she will develop a degree of detachment, turning down her sensitivity, for the initial contact with the beast. Then she will open up only as far as necessary or comfortable to detect and describe the symptoms.

A common discontent?

After years of working with horses, Lauren and I have felt a strong common undertone from many of her conversations with the animals. It was hard to define for the longest time, but the feeling was not pleasant. We started looking for greater clues about this unspoken message at each barn that we visited. A few, especially those where laughter and camaraderie were obvious, had no taint of the desperate silence, but in most business-like establishments the scent of the secret was very obvious. Our best attempts at defining the feeling were that the horses all knew that there had been an unspoken but understandable agreement between mankind and Horse (our generic term for all domestic equines) and that agreement had been changed.

And Horse did not break, alter or agree to the changes in the contract. Horse had become diminished in value; certainly not in terms of dollars, but as cherished companion and agreeable helpmate. In some barns the feeling was similar to dealing with disgruntled sweatshop labourers, in others there was a distinct atmosphere of slavery. The concern about buying and selling, being locked up or turned out on the barn operators schedule with a limited number of others whom they had not chosen as herd-mates, neutering or being bred at the owners timing and whim, harsh treatment and whipping for acting like horses, all of these things sat in the back of the attitudes of even the most talented and responsive champions.

Unfortunately, it is not advisable to speak to owners and barn operators about this undertone. It would possibly be taken as a direct criticism and may cause greater separation between the horse and its people. This could eliminate any possible future contact with that barn, and could put Lauren in the position of a potential disruptive influence. Thus she would not be able to speak for the animals again when her assistance could be helpful.

Spokesperson for animals
Because of the discovery of this underlying discontent, Lauren has adopted the title and purpose of spokesperson for the animals that she meets. Like any good advocate or talent agent, she attempts to foster better relations between the animals and their owners, and suggests more favourable conditions wherever appropriate. So now as a communicator, she uses her talents even more for the benefit of the animals than previously. This is not to be taken as a disparaging criticism of the owners, most of whom are doing their absolute best for their four footed friends, but occasionally a small change in the conditions or treatment of the animal can make life a little easier and will improve the fellowship between beast and person.

Word-of-mouth best advertisement

When she works at a boarding stable, Lauren always draws a crowd. As soon as she starts talking to horses, everyone in the stable gravitates to the event. Curiosity and skepticism are often the first drawing forces, but soon entertainment and genuine interest develop. The gathering throng soon emits laughter and questions, and many will run to bring in their horses to take part, if there is time available.

Lauren works with equal facility with other animals as well. Dogs, cats, rats, birds, lizards, mice, snakes and even more exotic creatures have told her of the conditions and events in their lives. (Caution: dogs and cats can and often do tell family secrets.) She has officiated at an open house at a "cats only" hospital, met with "fly-ball" and agility dog groups (where the hyper pets were literally "off the wall"), and she has lectured to the Guide Dog Users of Canada (where the dogs were incredibly calm, cool and confident). Lauren has appeared on TV on several occasions, picked winning horses at the track on the request of officiating organizations and worked with charities such as CARD (the Community Association for Riding for the Disabled). She has received very good write-ups in magazines and newspapers as a communicator and finder of missing pets. She has been a contributor to national magazines and papers, and has her own website (www.animaltalk.ca). Yet despite all of this coverage in the media, almost all of Lauren's clients have contacted her because of word-of-mouth recommendation.

Speaking to animals is a great joy in Lauren's life, and she feels that, through greater understanding, she can improve conditions for the creatures that she sees. But she is only one person. That is why she is wholly in favour of sharing her ability with others by training agreeable animal lovers to communicate as she does. She does not see other animal communicators as competition, but rather as a growing brotherhood or sisterhood of like-minded people concerned with the welfare of our nonhuman friends. Good work by any of her fellow communicators does not limit her potential, but raises the awareness of this valuable service to all pet owners. We wish you all success in the study of these following lessons and we hope that you will find delight in better appreciation of your animal friends.

Chapter One:

Starting Out

The dissenter is every human being at those moments of his life when he resigns momentarily from the herd and thinks for himself.

Archibald MacLeish

In practically every ancient heritage there are legends of people speaking to animals. On every continent there are stories of people living with animals, sharing wisdom with them and depending on them. Whether you consider Hindu legends, the stories of the Greek teacher, Aesop, the folklore of the North American First Nations people, Australian Aborigines or even the Bible (Eve and the serpent, Balaam and his ass), it would appear that by the shear weight of the common theme, there may be some possibility that communication between humans and animals did occur. There is a theory that homo sapiens, or their predecessors, communicated not only with animals but also between themselves in other than a spoken language. At that time probably all of the species exchanged survival information without dependence on words. Later, when spoken language evolved to the point where it served well enough to transmit thoughts and feelings, many of the rank and file lost the ability to use the old processes. Perhaps the shamen, witch doctors and medicine men retained and passed on the old knowledge for "calling" animals for the hunt, forecasting weather conditions, and other esoteric practices. As modern religious beliefs and "scientific" teachings destroyed the credibility of interspecies communication and similar intuitive abilities, the gifts disappeared or were driven underground. These feats are "unprovable" to the modern skeptic who refuses to accept the evidence of his own senses and suspects chicanery from all practitioners of these arts. But the ability that was once common has not been taken from our basic nature. I believe that in the genetic pattern present in every birth, there remains an innate ability to communicate without language. And it can be re-established in those who are prepared to do the work.

Animaltalk can lead to other discoveries

However, with modern students of the art, there is more to speaking with animals than the act of communicating in itself. With it comes the suspicion, and later the certainty, that there is a spiritual connection through all living things. You may find that many of the "scientific facts" about the workings of this world are only convenient fictions. Be prepared to have your world shaken. When the dust settles, you may discover that you still walk on familiar ground but you may tend to doubt things that "everyone knows" and embrace other beliefs that have been scoffed at. And I believe that your new world will be better for this.

I believe that we all have the ability to communicate nonverbally with animals through our intuition. How much of our intuition we are able to uncover depends on the individual. I say "uncover" rather than "develop" because I believe that we all have had the ability to sense things that cannot be heard, seen, touched or smelled with the physical senses, but somewhere along the line, we have convinced ourselves that "that is too weird." " We can't do that!" So the intuitive part of us has been buried under the desire to conform to "realistic logical thought" and "scientific proof". And this poor standard is based on the word of the authorities in our lives from our parents and kindergarten teachers through to the TV news analyst, medical and scientific naysayers and even the neighbour who says "Animal Communication? Totally wacko!" For some of us, the "uncovering" may require a good sized spade, with others, a bulldozer.

But it can be done. With practice and a true commitment, I believe anyone can learn to talk to animals as I do.

Animals understand more than we expect

So, how do we connect with animals through intuitive communication? First of all, you had better believe that animals understand more than we give them credit for. They can pick up information from our talk, from our attitude and from our feelings. So that side of the communication is relatively easy. Try it out if you haven't already. Speak to your horse or maybe a dog with intention and purpose. Believe that he or she can pick up your message. Try to erase any negativity in your voice, your stance and your feelings about the ability of the animal to understand. You may be surprised at the results; the attention that the animal gives you, as if enjoying the talk and wanting more, the apparent desire on the part of the creature to respond and return a message to you and perhaps a physical response in compliance to your wishes. (Your first assignment at the end of this chapter.)

Your intuition

Now the hardest part for most people is trying to receive a message directly from an animal. This is where intuition comes in. Through intuition, you can see without eyes, hear without ears, feel without touching and know without knowing how you know. Those are the four methods of communicating through intuition. Through Vision, through Hearing, through Feelings and through Knowing. We all use these gifts now, and we just have to fine-tune the sensitivity.

Close your eyes and picture this. You are standing in front of your refrigerator at home. You open the door. Where is the milk? You see it, don't you? That is the visionary gift. You did not use your eyes, but you saw the milk.

Now think. Who is your favorite singer? Which of this singer's songs do you like best? Can you hum along with it? That is hearing without using your ears.

During last winter's cold spell or in previous years, did you ever have to scrape off your car windshield without wearing gloves? Think about it. Can you recall how it felt? Do you know what it feels like to have your fingers tingle painfully from the cold? That is feeling without touch.

Have you ever had your phone ring and knew who was calling before you answered? That is knowing without knowing how.

We have these "extra-sensory abilities", but we ignore them. To communicate with animals, we must acknowledge these abilities, take them out of hiding, dust them off and consciously use them.

How animaltalk is received
When I talk to animals, I get the feelings from the body of the animal. I get pictures that are sent to me from the creature. I sometimes get sounds or words directly from the animal or how it actually heard them in the past. Or I just know what the animal knows and what it wants me to tell its human partner. And that is an instantaneous transmission. In a flash, I can get more information than I can say in ten minutes, or I may have to dig over and over to get the necessary info. That's why I can sometimes give a quick flowing reading, or on other occasions have a communication where I am silent for fairly long periods of time. The silence is not the usual thing for me though, as my husband will confirm.

Why animaltalk?
Now, why would anyone want to talk to animals? Ignoring the fact that they are often our good friends (and who else would we rather talk to?) anyone who has a horse will find many specific reasons to want an honest conversation with their mount. How about health? Before seeing the vet, or even with the vet present, wouldn't it be handy to be able

to describe the symptoms of a mysterious ailment as felt by the patient?

Perhaps your horse has a sudden change in behaviour. Is it because of pain in the body? An aching tooth? A pulled muscle? Hoof problems? A change in routine that upsets the animal? Or is it trying to tell you of a tack problem, a poorly fitted saddle or bit?

Perhaps closer to your hearts: what if the horse is particularly slow in its schooling? Maybe something has been missed in the instruction. Maybe the horse has a peculiar phobia due to a past experience that you have no other way of knowing about.

Wouldn't it be great if your horse could brief you on specific details of incidents occurring while you are away from the stable, or warn you of the misdeeds of other animals or people in your absence? Granted, the horse's testimony would not stand up in a court of law, but forewarned is forearmed.

Animaltalk is like imagination

There is a much-used quote from Albert Einstein that says: "Imagination is more important than knowledge." He also said "When I examine myself and my methods of thought, I come to the conclusion that the gift of fantasy has meant more to me than my talent to absorb positive knowledge."

Obviously "positive" or proven knowledge is limited, while imagination knows no limits.

Animal communication is just like imagination. For you it is new and far beyond any positive knowledge in your memory. So we must start in the imagination, which opens up the world of "what if". What if we can really understand animals in words, feelings, emotions and every one of the senses? Hold that as a basic premise. For a while drop the impulse that says "But I can't".

Try it yourself
Forgetting animals for a moment, let's try to get a feeling of what this type of telepathic communication is like.

Think of a problem, a question, a decision that is facing you for which you would like to have another qualified person's opinion.

Think of a person whose opinion you respect. No restrictions. A contemporary, someone from the past, alive or dead, but someone whom you respect for their understanding and knowledge.

Using any one or all of the gifts: visualization, feeling, just knowing, hearing the voice of that person, feel your self in contact with him or her. You may imagine that you are that person (just as an actor takes on a role), or you may feel that you are in their presence. Ask the question and sense the answer that your friend would give you.

When you can be sure that "Yeah. That is the answer that he (she) would give," then you have it. This is the same feeling or knowing that you will get when you talk to animals.

Bypass the "reality filter"

When we try to recapture the degree of sensitivity that we were born with and trained out of, it is very common to have that inhibiting rationality, the "get real" filter in our brain, tell us that it is only imagination. That is its job. Fearful conformation to the "real" world of pseudo science. Fearful that you might not fit in if you seem to be different. They might lock you away, figuratively or physically. Your job now is to override that inhibition. Don't worry if you can't turn it off like a switch. It didn't develop overnight. It took years of training and you probably used that "Imagination" label to squelch many intuitive incidents every day of your adult life. (Most often, people allow their conscious mind to accept intuitive thoughts or feelings to direct their actions only in the case of "fight or flight". In the emergency mode the conscious mind will accept anything, even advice from the all-knowing subconscious, just to get out of the emergency.)

So learning to pay attention to thoughts or feelings that you have habitually labeled "imagination" (as if it were a bad thing) is not an easy task. It will be easier with a group of people who are all trying to do the same thing, but for some, even there it will be a big stretch. Be patient. And be aware, you may revert to that old status that labels what we will do here as preposterous. It will take work, consistent usage and constant permission from yourself or ego to use your "new" skills, before it becomes a part of your life.

> *You cannot see anything that you do*
> *not first contemplate as a reality.*
> Ramtha

Obviously, a big priority of this course is to learn to suspend your disbelief. Allow your mind to accept what you have called "imagination" as a reality. Here are some things that you may want to accept in this world of "what if", at least for now. Proof or knowing may (or may not) come later.

- There is a collective mind that contains all experiences of all sentient beings that exist or did exist.
- These experiences and this knowledge are connected to you now through that collective mind. As you contribute to its content, its content is available to you.
- Animals, because they have life, can think, can react to their environment, are sentient beings just as (most?) people are. They are part of that collective pool of information (and in reality they may be much more connected to it).
- You can bridge that gap between your thoughts and an animal's thoughts through that collective consciousness.

Now if you are still with us and you haven't been deafened by all of those alarms and warnings ringing through your head (Wow, wow, twilight zone! Houston, we have a problem! Watch out, the men with white coats and a long sleeved jacket are about to break in!) - let's get on with it.

Assignment 1-1

Go back to the second page of this chapter to the paragraph beginning "So, how do we connect with animals through intuitive communication?" Follow the instructions about sending a message to an animal. Some helpful tips:

- Use a pet or horse that you know very well.

- Relax and bring your breathing under control.

- Find a very good feature or trait of the animal that you can concentrate on.

- Send a complimentary thought to the animal about that feature.

- Feel that the animal has received that feeling of admiration.

- When you feel that your four-legged friend has reacted to this warm thought, send your planned message with intention and confidence.

- Observe and record any responses (or feelings that you may receive).

Assignment 1-2

Take stock of your acknowledged intuitive abilities now. With a piece of paper and a pen and answer the following questions with an appropriate answer of "never", "once or twice", or "often".

1. I can walk into a room and sense if there is a conflict between the people in it.

2. When I meet people I can recognize and respond (in like manner or in opposition) to their emotions of
   a. despondency
   b. pain
   c. anger
   d. boredom
   e. complacency
   f. cheerfulness
   g. enthusiasm
   (Score on the strongest empathetic emotion only, and note which it is.)

3. I can tell if someone is lying or has a hidden agenda
   a. by the tone of their voice.
   b. by the look on their face
   c. by simply sensing (feeling) it
   d. by a knowing that comes for no logical reason.
   (Score on only one of the above, and note which alternate is strongest.)

4. I can complete sentences for my friends or partner if they pause.

5. If a friend is ill or injured, I can sense (feel) the ailment in my own body.

6. I do empathize with a disappointed friend.

7. I am joyful over the success of a friend.

8. When trying to emphasize an important comment to others, I often feel the urge to touch them (hand on shoulder, arm, hands, etc.)

9. I can "get the picture" when a friend describes a person, place or thing.

10. I can visualize how a colour from a swatch would look on a wall or large object.

11. When getting directions, I can draw a map in my mind.

12. I can pick out an unusual accessory that will enhance a dress or suit without having to try it on first.

13. By looking at a fruit or a prepared meal, I can sense how it will taste.

14. At the first taste of a prepared dish I can tell what would improve the recipe.

15. I can tell who is calling before I answer the phone.

16. I can recognize the emotional state of the caller by just hearing them say hello.

17. I am certain of my resulting success in sports before
    a. the golf club hits the ball.
    b. the basketball leaves my hands.
    c. my racquet hits the tennis ball.
    d. the bowling ball leaves my hand.
    e. or similar certainty in other activities.
    (Score on one activity only.)

18. When I have a thought about something out of the ordinary, in very short order someone else brings up the subject.

19. When I need to make a decision and am looking for an answer, it appears to me in usual and sometimes ridiculously unrelated places.

20. I already think that I know how I will score on this or similar tests.

Score zero for every "never", one point for every "once or twice", and three points for every "often". Possible scores range from zero to 60, but we will not classify your position now. This is mainly for your own information and you may wish to revisit this test a week or so after you have completed the course to see if there are any changes.

Your answers will assist in assessing your communication strengths. High scores on question 1, 3c, 5, 6, 7,and 8 indicate a high degree of the feeling gift. High scores on questions 3b, 9,10,11, and 12 show a visionary gift. Highs on 3a, 4, 16 and 19 suggest a strong intuitive nature that relates to sounds. Highs on 3d, 12, 13, 14, 15 and 20 indicate a prophetic personality. Your answer to question 2 might provide a clue to your habitual emotional state or it may indicate a past experience of high emotion that still affects your decisions today. Question 17 is there to assist you in recalling the feeling of being "in the zone", as they say in sports. This is a fleeting but real contact with spiritual communication when all four gifts apply. Ideally, if we were able to live "in the zone", animal (and interpersonal) communication would be a cinch.

CHAPTER TWO.

SENSITIVITY

The most beautiful thing we can experience is the mystical. It is the source of all true art and science.

Albert Einstein

As you can imagine, animal communication requires a certain higher order of sensitivity. If you are to receive messages from the animals that do not rely on exact physically observable prompts such as vocalizing in a known language, posturing in accordance with an understandable sign language, or physical nudging to point you in a desired direction, then you must be prepared to embrace the idea that you are capable of greater sensitivity that will allow you to see beyond the current use of your eyes, hear beyond the sounds heard by your ears and feel without physical contact. You must also be ready to "know" things without having to analyze how you know them. Now this may sound extremely exotic or beyond your own capabilities, but I do not believe that it is. I believe that these abilities lie dormant in all of us blocked away in a dusty old area of our minds and brains surrounded by posted warnings, "Keep out. Trespassers will be ostracized and ridiculed!"

Refusing the evidence of our senses
There is a persistent legend about the first arrival of Columbus on the islands of the Caribbean. It is said that the natives of the island were unable to see the Nina, Pinta and Santa Maria as they approached the shore, because

they were not aware that such large floating islands could exist. Having nothing comparable to relate to these objects, the Indians, as Columbus named them, refused to see the ships until men (who were recognizable) strode out of the water and onto dry land.

Whether or not this story is true, our so-called psychic gifts can remain invisible and hidden in a similar manner. The idea of discovering information through means other than our accepted physical senses is so foreign to us that we block anything that we receive through our other sensitivities, and label it imagination (as if that was a bad word), illusion or wishful thinking. We even prefer to call psychic messages mental disturbances or chemical flashbacks, rather than give credence to them. In rare moments of confidence in our abilities we might accept these clues and file them under "hunches" or "gut feelings".

Now where do you suppose these hunches and gut feelings come from? True, it would appear that many could be attributed to information stored in our subconscious mind and not readily accessible to conscious thought until stimulated by a key word or situation. But that does not explain the novel ideas that come from thinking "outside of the box"; the brilliant forward-thinking approaches that revolutionize our lives. These innovations may not come from past experience or subconsciously buried information. Could it be that we can draw from a collective intelligence that is not limited to past experience and the past-to-present sequence of time? Could we be sensitive to power and intelligence that transmits to us in a manner beyond the limits of our physical senses?

Human senses are limited

If we examine how we see, hear, smell and feel it should not be so surprising to imagine that we have other avenues of receiving input. The eye does not "see" on its own. Light passing through the lens of the eye is focused on the retina through muscular shaping of the lens. The retina does not reproduce an image, but through some electro-chemical process transmits a message to the brain, which decodes the incoming data to present an impression to the thinking mind. Similarly the ear concentrates incoming vibration causing movement of a diaphragm, which activates tiny bones to produce sensible movement. The feeling of this movement is transmitted to the brain for analysis. The same applies to touch and smell. Nothing is seen, heard, smelled or felt until the brain receives a stimulus. This is all very wonderful and amazing, but the human senses are extremely limited.

We do not hear the low frequency sounds that can be produced and sensed by elephants and whales. We do not hear high-pitched sounds that dogs or bats can hear. We cannot see in the ultravoilet or infrared ranges, but we have instruments that tell us that those invisible light fields exist. We cannot smell as well as hounds or snakes. Our sense of feeling is not as sharp as many animals and birds that seem to use magnetic and gravitational forces as guides for migration. In fact our physical senses are so limited, even in comparison to our animal friends, one might wonder how mankind got to be so dominant.

Mankind tried to rule nature

Yet we did leap ahead of other species to such an extent that we have restructured our environment. That used to be one of the major brags of our scientific community: "Animals may adapt to their environment, but man shapes his environment to suit himself". I am afraid that we got carried away too far with that. With overpopulation, pollution, destruction of the animal habitat (and our own), mismanagement of our natural resources and destructive chemical practices, we threaten even our own existence.

We have come far from the hunter/gatherer who roamed the forests and plains of untamed nature and left very little trace of his passing on the green landscape.

When Europeans first came to North America, their behaviour contrasted obviously and radically from the largely nomadic Native Americans. The Indians lived in harmony with nature. The white man's attitude toward the land was "destroy the forests and build farms, roads and towns". Toward the animal kingdom that sustained the red man, the newcomers took the attitude of "slaughter it and use it for food, use its furs for commerce or if it has no commercial value, just slaughter it anyway so that it won't eat our crops or bother us in any way". We had come a long way. But as we moved toward manmade communities and a social structure that aimed at totally abandoning the natural setting, perhaps we left special survival talents behind.

Many of the aboriginals of the Americas, Australia and the Pacific Rim countries bewildered the more scientifically enlightened explorers by living at peace with their environment and being able to harvest crops and animal protein without the need for land claims and ownership so dear to the European attitude. They had great knowledge of the natural ways of native animals and vegetation. They knew when to plant, when to hunt and when to move on.

The Shaman
Most tribes had a spiritual leader, a shaman, who ranked just below the chief. The shaman gave advice, healed the sick, communed with nature and performed rituals to assure good harvests and good hunting. Many of these advisors were considered "callers" who could summon game for the hunt and tell the tribe where and when to go to intercept their prey. Their wisdom was passed on to selected acolytes in a process that took many years and many trials.

In the early days when the people traveled in single families or in small packs, probably there were many, if not all of the tribesmen, who could perform such wonders, but as tribes grew larger, specialization increased the tribe productivity and ensured greater survival. Some were better hunters, stalkers, gatherers, toolmakers and so on. They practiced and refined their crafts for the benefit of the whole tribe, knowing that in trade for their specialty they would be able to rely on others for the things and services in which they were not proficient. And so the esoteric skills were left to the best in that field too. Thus was born the shaman who preserved and performed the essential spiritual skills for the whole tribe.

Influence of "race mind"

I believe that back in our own prehistory a similar process took place. In fact, I believe that to a certain degree it is still taking place with every child born into our modern world. At birth, we are not much different from prehistoric babies. We probably have much of the same potentials. The difference is that today's children are influenced immediately (either from conception or from birth) by the field of beliefs and information prevalent in our current social structure. (You may call this field "race mind", or "social awareness".) Being totally immersed in this field of great influence, a child quickly develops a sense of limitations and what is acceptable in its immediate arena of interaction with its peers and elders, from whom it receives sustenance and protection.

So today's youngsters easily and fearlessly accept electronics and computers, jet travel and satellites, and all other very recent wonders that we have had to struggle with, but they, just as quickly, learn what they are not supposed to be able to do: like see and feel auras, sense future stresses, see non-physical things and talk to animals. Being born in the backcountry in South America, away from "civilized" influence and to a family that could tolerate aberrant talents, gave me a great advantage.

Unlearning

For you to "learn" to talk to animals, you must be prepared to unlearn limitations that you have carried with you for a lifetime. Then you must practice seeing, hearing and feeling beyond those limitations. This requires opening to greater, extraordinary sensitivity. I use the word extraordinary with purpose. I do not mean unusual or odd, but beyond the ordinary sensitivity that we apply in our daily lives in the so-called "real" world. Bear in mind that if you plan to do something differently, beyond the limits of your usual activities, you must change your mind. Open to new possibilities. Be prepared to do something different, trusting that you will be okay, and you will, in fact, discover that you are capable of observing your whole world differently. I know that many of us are afraid of change, or at least are super-cautious about radical ideas that may force us to change our usual routine. For you I simply ask that you "play" along with us and enter into the realm of "what if - ?" Use your imagination and drop your doubts and convictions that this is impossible. Humour me and imagine "What if all this were possible and I could pick up thoughts and feelings from animals and even other people without any outward and possibly faked sights and sounds?"

Social conditioning

I believe that as very young children, we had and probably used avenues of communication that did not rely only on physically displayed signs or sounds. But even before we entered kindergarten, we were trained to deal with the physical world in rather sophisticated ways that required our full attention and adherence to the social norm. We were taught to dress in an acceptable fashion, to behave at the dinner table, to speak only in a certain way to adults, to get along with our siblings, to stay out of trouble. All of these things required training and retraining until we could get through the day without too many reprimands. Then we went to school. We learned how to sit, how to stand in line, how to be quiet when we wanted to shout, how to hold up a hand when - - you know, how to get passable grades by feeding information back to the teacher in the way that was expected, how to please authority figures or at least be invisible to them. And that was only the beginning. Coupled with this social control was the inflexible rule that did not tolerate individuality. To be different was punishable. Creativity had no place in the education process. Conform or be scorned was the rule that was actively and passively enforced. Today, pressures from the home, the workplace, social peers, government, church and almost every facet of our lives have set rules for us that must be learned and obeyed. That we have survived and function in this environment is proof that we have mastered much and have adapted to the game.

We haven't lost our innate abilities to communicate through "telepathic" means. We have just buried them. And how we buried them! As babies, when we discovered that our nonverbal, nonphysical attempts at communication failed to reach our parents, we resorted to loud sounds and acting out. That worked to a degree, and we discovered that the louder the sounds, the more attention we got. Growing up we learned to fortify that adage. We quickly learned that the greater the volume and force, the quicker the response. It carried over to our childhood and adult rules of life. "The squeaky wheel gets the grease." "Nothing succeeds like excess." So not only have we give up any attempts at nonvocal communication, but also we have been conditioned to expect greater response when exaggerated emotional vocals are played.

So how do we recover the very delicate nuances of nonvocal communication? It is a long way back from this noisy world where we know that many of the forceful messages that we are getting have been falsified for effect. (Observe the hyper kid in the toy store, or the politician who hammers home the same old lie often enough that it becomes an accepted truth.)

Trust your own observations
We must drop the socially conditioned responses that forcefully tell us what is possible and what is not, and judge our present situations solely on the evidence of our intelligent senses. Simple, but difficult. Because of the conditioned dominance of our left brain, it will be probably impossible for us to block out this prejudiced thinking (that is what it is) for more than a few moments at first. And in order to obtain even those few precious moments, we require a bit of conditioning. This will take the form of a guided meditation designed to "ground" you in the present and open you to other possibilities.

Chapter Three.

Opening Meditation

In meditation, effort must be applied in a direction opposite to what we are used to. Our "effort" must be to relax ever more deeply. We must ultimately release the tension from both our muscles and our thoughts. When we relax so deeply that we are able to internalize the energy of the senses, the mind becomes focused and a tremendous flow of energy is awakened. ... Meditation is a continuous process, and can be said to have three stages: relaxation, interiorization, and expansion.

John Novak, Lessons in Meditation.

This particular meditation borrows from the Kundalini Yoga tradition dealing with power centers in the human body called chakras. For those who have not previously been exposed to the concept of energy flow through the body and these control centers or chakras, here is a very brief explanation.

Chakra is a Sanskrit word meaning "wheel". There are seven chakras in the human body and they are visualized as spinning vortexes of colour and energy located to match seven main nerve ganglia that branch off from the spine. Each chakra relates to a specific area of the body, a physical and/or psychological function, and a spiritual and emotional state. Each chakra has its own colour and responsive musical tone. An imbalance at any of the chakras may indicate a physical and psychological problem, perhaps due to the blockage of energy through the body. The blockages may cause the symptoms, or the ailment may cause the blockages. Either way the evaluation of energy flow through the chakras is an excellent way to diagnose overall health, and by balancing the chakras one can induce health improvements in the body.

The following meditation will provide information about the names, location, function and purpose of each of the chakras. I expect that it will be a pleasant and more effective way of introducing the characteristics of the energy flow than writing in detail here.

One person may read the script for this meditation for another or for a group, or it may be prerecorded on an audiotape for access at any time. The reader should speak quietly, clearly and with a friendly, slow but positive voice as if reading something of importance to a best friend (and I hope that is the case). Read the script in full and exactly as written. Pause for the duration of one breath between sentences and two breaths between paragraphs. Instructions to the reader will be given within sharp brackets < > and are not to be spoken aloud. Of course, if this is to be a "live" reading, the reader must limit his or her participation to that of reader only. Do not attempt to go on this trip with your audience. Hopefully, the roles will be reversed later.

So let the fun begin!

Meditation Script

<Speak slowly, clearly and firmly>

First get comfortably seated in a position that you can maintain for about twenty minutes. Both feet comfortably on the floor. Do not cross your legs, but place your feet side by side, not touching. Place your hands on your thighs, comfortable with the fingers in a relaxed position, not held stiffly straight. Relax your neck and shoulders. Do a few neck rolls if you wish, and then looking straight ahead, relax and close your eyes, simply to remove any visual distraction.

Relax your shoulders. Raise them up to your ears, move them about and then let them drop into a lowered relaxed position. Feel your body. Any tension? Any discomfort? Move or adjust your position to get as relaxed as possible. Let the muscles go loose and limp. Feel your weight against the chair and the floor. Very comfortable.

This is your time. Time set aside for your own good. Nothing else matters right now. Everything else can wait. You have no "present time problems". This session is for you. You have earned it. Any outside noises that you may hear will only serve to remind you that the world is continuing on its own, as it should and it needs no help from you. The noises will serve to relax you further, knowing that you are in a safe environment, taking this time for yourself, and you will relax deeper and deeper with each noise that you hear.

Now pay attention to your breathing. I want you to feel the air that you are breathing in, feel the breath that you are breathing out. Relaxed breathing. Not timed or set in any specific pattern, just relaxed and natural. Feel your lungs fill and empty. Visualize good clean health and energy in the air that you inhale and your exhale expels all tension and discord from your body. Picture that.

Inhaling health and energy. Exhaling tension and any discomfort.

Move your attention to your feet. Feel them pressing against the floor. Feel the soles of your feet and the pressure of the floor on them.

Imagine that little roots are starting to sprout from the bottoms of your feet. The roots grow and push right through the floor that your feet are resting on. They continue to grow and extend downward through the floor and the foundations of this building, right into the soil below. The roots grow and grow, extending downwards, deep into the earth. With them, the roots are taking any tensions and unhealthy substances from your body and expelling them into the earth where this waste is accepted hungrily as humus and a good fertilizer that the earth needs.

Now the roots are drawing energy of the earth in the form of a clear liquid substance and this substance rises up through the roots through your feet, your legs, to your body. As this energy enters your body at your first chakra, located at the base of your spine it is red in colour. Red the colour of this root chakra, the energy center that represents your needs for survival in this physical body. The red energy flushes through the root chakra taking away fear and allowing you to feel completely in the here and now.

As the energy flows upward into your body, the colour changes to orange as it reaches your sacral chakra, located about two inches below your navel. This chakra deals with needs, emotions, sexuality and intimacy. The energy rises through the second chakra balancing your openness to other people.

The flow from the earth now turns yellow as it reaches your third, solar plexus chakra. This is your identity center. It deals with your feelings of self-esteem and self-image. Passing through, the earth energy allows you to be decisive and in control.

Next the body is filled with energy to the heart chakra and the energy becomes emerald green. The heart chakra is in the middle of your body at the level of your heart, and it is about love and compassion. The energy opens you to great affinity for others and harmonious relationships.

Continuing, the earth energy flows further up to the throat chakra turning blue as it reaches your throat. This is the center of communication. It encompasses your throat, your lungs, your bronchial tubes, your thyroid, your mouth, your ears and nose. This is your speech area. The blue energy releases your fear of criticism and helps you obtain a clearer more resonant voice.

The energy turns to an indigo colour as it reaches your third eye chakra in the center of your forehead just above the eyebrows. This is the center for insight, imagination and visualization, intuition and psychic ability. The energy cleans this chakra allowing you to use imagination, even fantasy without losing your reality. It allows you to evaluate where you are on your path to spirituality and healing.

The energy flow turns violet in colour as it reaches the last chakra, the crown, on the top of the head. This is your connection to harmony with the universe, God or spirit, through your higher self. It goes beyond language, time and space. Love, peace of mind and connection to spirit are balanced with the need to own and maintain your physical body.

The earth energy continues to enter your body, and now a small opening begins to appear at the crown chakra. It grows to the size of a quarter, a 25-cent piece, and the energy showers out of your body into the air. It cascades down your body and sprays outward to the ground, blessing it and cleansing an area around you.

Now clear white vinelike tendrils start to sprout through the hole at the crown area of your head and grow upward, reaching, straining higher and higher until they reach beyond earth's atmosphere and shoot to the stars and beyond. New crystal clear energies start to flow downward from the heavens through these tendrils, down, down to your body and through your body to the ground beneath. This heavenly energy enriches the earth and you are the conduit. You are the connection between earth and infinity and you are the path, the way, through which the physical relates to the infinite. You are in a physical body, but you are in contact with nature and the infinite. You are the bridge. You have all of the resources of earthly knowledge and the unknowable. You are not limited.

Please relax as you are. Let your mind drift as it wishes. Enjoy the moment and the feelings. Give yourself another minute of peace. <Pause for a full minute.> As you slowly become aware of your breathing again, and you feel your connections to the earth and the sky fade into invisibility, in your own time, come back to this room and keeping your eyes closed for another moment put together your experiences from this meditation. When you are ready, open your eyes, and taking your writing materials make notes on your experience. Then, stretch and feel free to move about the room, if you wish.

Chapter Four.
Feeling the Aura

*The significant problems we face cannot
be solved at the same level of thinking
we were at when we created them.*

Albert Einstein

Almost everyone has heard of auras, but who really knows what they are? I have heard some people proclaim (erroneously) that the aura is the real you. Although I applaud the recognition that we are more than our physical bodies, I also insist that we are more than our auras. The real you, the real I, is not limited in space or time and is not necessarily confined in position or influence to a body or the slightly larger aura.

In his book "Space –Time and Beyond", Dr Fred Alan Wolfe, a theoretical physicist, says that an aura is "a glow that is not light, that cannot be perceived by the normal senses, but that is sometimes seen by people in altered states of consciousness". He adds "I suspect that auras have something to do with the quantum wave function that not only exists in our minds and brains but also exists in our space–time" (the physical universe). "The observer who is sensitive … could know about it through one of his own five senses. For example, he might smell something, or see a glow of purple. Or he might hear voices or feel a foreign presence."

The Aura is an indicator

So, if the aura is not light (at least not light that can be scientifically observed and measured) and it is not the outline of the "real self", just what is it? I suggest that the aura is a field about the body that is influenced by physical, mental and emotional energy and changes of energy. Note that I say the aura is not the energy itself, but a field that is affected by the energy in such a fashion that the field can be read as indicator of the physical, mental or emotional state. Just as the red colour of a hot toaster element is not the heat itself but indicates the presence of the heat, so the auric field is an indicator. That auras are affected by physical condition and energy is evidenced by the use aura-reading in health diagnosis. Mental and emotional energy changes account for more spectacular aura fireworks as observed by experienced readers.

The aura surrounds the body and extends out from it with varying degrees of intensity. Most aura readers describe it as a roughly egg-shaped field, thicker at the top, the head level, and tapering down to the feet. I believe that this is due to the degrees of mental and emotional activity affecting specific parts of the body. We observe and communicate more with the head than any other part of the body, and the legs and feet are usually limited to physical activity. If we saw, heard, talked and created artistically with our feet, perhaps the aura would be reversed.

Next to the head in aura intensity are the hands. With these we provide for and feed ourselves, work, create, and often communicate (to a degree that often varies with parental example and national origin). So it is not surprising, considering the multitude of intelligent ways that we employ our hands, to discover that the energy level and the resulting aura around the hands is quite powerful. Through practice and developed sensitivity, the auric energy around the hands can be used for reading and healing.

Enough background theory, let's do something.

Prepare by relaxing
Earlier, I discussed the four basic gifts of communication. One would think that the study of auras would be most applicable to the clairvoyant gift, seeing without eyes, but we will pass over that approach so that we may develop a tactile skill that will be of assistance to us in relaxing and tuning into a receptive state that will help us in all of the practical exercises that follow this chapter.

Because we will be working with sensitivity, I want you to be fully relaxed. Seat yourself in a comfortable chair, feet on the floor in front of you. Relax as you would before entering meditation. Use all that good stuff involving release of body tension and mental busyness. A couple of slow breaths now, then let us begin.

Feeling your hand energy
Place the palms of your hands together and rub them lightly to enhance their sensitivity.

Now, position your hands, palms facing each other, about eighteen inches apart, then slowly move them together to about two inches apart.

Do you feel anything? A warmth, a feeling of compression, an unusual feeling of greater density or perhaps an awareness of each of the hands felt on the other?

What you are feeling is the aura.

Move your hands apart very slowly until you can no longer sense that feeling.

Repeat moving them together and apart until you can feel the buildup of intensity and can easily sense the range of the feeling.

There, you have felt an aura. On to the next step.

Tactile sensing ability
The energy of the hands has a wonderful healing quality. Perhaps that is why we instinctively place a hand over a scratch or a burn to start the healing process as soon as we are injured. We will now combine the tactile sensing ability with the healing quality of the hands.

For this exercise, you will need to pair up with another person. Have your partner sit in a chair that allows you easy access to his or her head as you stand behind the chair.

Standing behind your partner, again relax. Take a couple of calming breaths.

Rub your hands together again as you did in the previous exercise. Now move your hands to a position so that the palms are about one inch away from your partner's head. Just where on the head doesn't matter right now, because you are going to allow your hands to roam all around it.

Do you feel the aura? If you are not sure, move your palms away slightly and then closer again until you are in touch with the energy.

Healing the aura
Now explore your partner's head aura, letting your palms slowly circle the head, staying about one inch away. Do you notice any cold spots? You might find one or more areas where it feels as if a cool breeze is hitting the palm of your hand. (To help you concentrate on the sensitivity of your hands, you may wish to close your eyes when exploring areas of variation.) These cool spots are breaks or weak areas in your partner's aura. When you locate one, slowly move your hand in a small circle around the cool spot until it disappears.

You have healed a break in your partner's aura by using the healing power of your hands. The energy from your hand has supplemented your partner's and has assisted to remold the aura over the opening. Repeat the procedure until you can find no more breaks.

On rare occasions, you might not be able to find any cool spots. Don't feel disappointed. It may mean that your partner is too healthy and at peace at the moment. In this state, he or she is of little value to you as a subject for the healing exercise, so get another partner (only for this practice session, of course).

Personal aura cleansing

Now that you are an expert on the tactile art of sensing auras and have proven the healing power of your hands, let's learn something that you can do for yourself, any time or any where, to help you heal your own aura and quickly reach a calmer, more meditative state.

1. Lightly rub your hands together or shake them briskly to heighten the sensitivity and charge your hands with energy. (With hands limp and relaxed, flick your wrists as if you are shaking water from your fingertips.)

2. Place your fingertips over and about an inch away from the center of your forehead (the "third eye" area) with palms extended across the forehead.

3. Move the palms down over your face to a position at the throat area.

4. Shake your hands briskly again to restore the energy.

5. Return your hands to the starting position: finger tips over the third eye.

6. This time, keeping your finger tips close together, move your hands up, over and to the back of your head, only an inch or so from it, following the shape of your head all the way around to the back of your neck (where the palms will be facing the pituitary gland). Continue in a smooth flowing motion, bringing your hands around each side of your neck and meeting again at the front of your throat (at the thyroid gland).

7. Repeat the whole procedure accompanying the physical act of aura cleansing with a positive affirmation (aloud or silent) as follows:

   "In the name of (*), I cleanse my body and I cleanse my soul."

   At (*) insert a term of special significance to you such as "God", "Universal Mind" "Love", "Prosperity", "Strength", "Health", "Mom's Apple Pie" or any other term that appropriately symbolizes the energy that you would like to hold in your thoughts.

8. Repeat until you feel a warmth or a tingling shiver.

The cleansing process is complete when you feel the "quickening of the flesh" (light shivers), a warm relaxed feeling or the sense that you are fully "centered" in the present, undisturbed by outside influences.

This process is recommended before and after important actions to compact your energy for the task at hand before meditation or simply to relieve stress. I introduce it here so that you may use it to assist you in preparing for any of the other practical exercises to follow.

Chapter Five.

No Present Time Problems

The only reason for time is so that everything doesn't happen at once.

Albert Einstein

I am sometimes amazed that there are people who cannot grasp the concept of "no present time problems". They seem to put great pride and ownership on their problems and refuse to consider putting them aside until they are solved or turn into a disabling ailment that gives them an excuse to give up. And that is not an exaggeration. Personal problems occupy most of the thoughts of some people during their waking hours and even disturb their sleep. Problems can be the main topic of conversation with family, friends and even strangers. Problems often are the strongest form of identity that these brooders relate to, affecting not only their own outlook but also others who deal with them on a daily basis. How many times have you thought when you recognize a constantly obsessed person heading your way, "Oh no! Here comes so-and-so. Where can I hide? I don't want to hear about his problem with (whatever) today"? These people stake a claim on their problems and emphatically own them by branding the situation with a first person possessive modifier. They speak about "my problems with my spouse", "my indigestion", "my debts", "my weight problem" or "my cancer", making it special, different from the generic brand, and defying anyone from attempting to separate them from the problem.

Constant ownership of a problem gives it energy, creative energy. Thought creates, and obsessive thought, allowing no alternatives, creates quickly. Constant attention to and ownership of a problem results in elevating that situation to a position of overwhelming importance and irresistible power. A child, who for some inexcusable reason, is called "stupid" or "dumb" by his siblings, peers, or some adult authority figure may try at first to disprove that designation. But if his tormentors persist long enough, he may accept the role of "dummy" and stop trying to excel or show origination. He may withdraw from any intellectual effort and, taking the word 'dumb" literally, retreat into a sullen non-communicative state.

Similarly and tragically, it did not surprise me when Terry Fox, an undisputed Canadian hero, succumbed to cancer during his cross-Canada run. He had just covered half of the incredible journey dwelling on the severity of the disease, speaking of it at every stop and interview, and thinking of the loss of his limb, consciously or not, on every painful and exhausting step of the way. Thought and attention have great powers to create.

Break the problem pattern
On the reverse side, this "thought creates" principle is part of the curative power of laughter, meditation, prayer and enjoyable creative hobbies. All of these things break the pattern of obsessive thought. Laughter spontaneously convulses us and breaks the hold of worry. It produces "feel-good" chemicals in the body that actively combat the solidification of the negative emotions. Prayer,

likewise, breaks the negative pattern, at least for a few moments and occasionally with a degree of permanence. In my definition of prayer, the supplicant contemplates his concept of the infinite God, Mind, Spirit or whatever creative energy best suits his religious or non-religious belief. He then considers his or her own relationship to that Infinite Source. This will put the person out of contact with the problem and the responsibility for that problem for at least a short period of time. The result is a definite break in the feeling of loss and negativity. The same applies to meditation and a constructive hobby. They provide a healing respite from the destructive power, the negative creativity, of a troubled mind.

Meditation, like the other positive recreations mentioned above, can move your thoughts and feelings into a state much like an "out-of-body" experience. It is very similar to what athletes call being "in the zone". It is the feeling of knowing that your drive is going straight down the fairway, before the club contacts the ball, the basketball is going through the net without touching the rim, before it leaves your fingertips, that you will score a strike, before the bowling ball leaves your hand. It is an exalted, timeless feeling that you are more than your body, that you control a greater space than your body. No wonder it is healing!

One of the beauties of reaching this state is that once discovered and experienced, it can be recaptured. Simply being still and recalling the feeling, you may find yourself moving, mentally and emotionally, back to that ethereal belief that things are just fine in your world.

Again, practice helps tremendously, as does the cleansing technique described in the previous chapter on feeling the aura. The sooner that you can feel that you are "in your zone" and declare to yourself that this is your trouble-free base line, the sooner you can get on with your animal contact (or any other extra sensory procedure that you may have learned). This is the experience that will assist you in gaining confidence in your own sensitivities that may allow you to receive the very delicate impulses of unspoken communication.

The first step is to be able to acknowledge that you are in a state in which you have "no present time problems". Please understand that when I want you to have no present time problems, I do not expect you to solve any or all of the dilemmas that you think about in your daily experience. I simply want you to define the term "present time" as a short period today that you have set aside for a specific purpose. We are not going to try to solve any problems. You can keep them. Just acknowledge that they will not change in the next hour or two, and it won't matter if you don't think of them while you are taking a break and doing something totally unrelated to them. Ignoring your problems for a short period of time will not make you a bad person, and it might even be therapeutic.

Release brain clutter
The real key to meditation is to think of one thing only in a relaxed and content state. And that falls into the category of things that are simple but not necessarily easy. To be successful at picking up impressions so delicate

as "hearing without ears", "seeing without eyes", feeling without touching" and "knowing without knowing how", you must at least get rid of the usual brain clutter that we have accepted as normal in our busy world.

The first step is to consciously relax and take stock of what is on your mind at the moment. Take a few slow breaths, focusing on your body and easing any tense muscles. Then think of what thoughts are going through your mind. Realize that this is a time that you have set aside for yourself, for your own benefit and growth. It is a well-deserved and valuable thing that you are engaged in now. The problems and uncompleted actions with which your mind has been nagging you do not matter for this moment in time. Sure, maybe your mind is telling you that you have to meet with someone, have to do some thing, someone is counting on you, and you have obligations. None of this is a present time problem. Letting go for a short while is not going to do any harm. This is your time. Here and now is the only important thing in your life right now. All else can wait, and will be handled in due time.

So let those bothersome thoughts come to your mind, acknowledge that you are not going to do anything about them right now, and let them go. Do not expand on any thought or try to find a solution. Just acknowledge the thought and let it go. The same with the next thought, and the next, until the jumbled torrent of problems smoothes out to a slow trickle and you feel more relaxed about having this moment for yourself. If there are any thoughts or perceived obligations that will not go away, or if you are afraid that you will forget them, write a few key words about them on a piece of paper so you can pick up where you left off later. Then fold the paper over and put it aside for now. This time is reserved for your Animaltalk practice session only. In this present time there are no problems.

Chapter Six.

Reading Animal Pictures

To be loved by a horse or any animal, should fill us with awe - for we have not deserved it.

Marion C Garretty

Although many people might consider that remote reading of animals is probably much more difficult than being in close proximity and having the opportunity to physically touch the animals, we will start your Animaltalk experience with remote picture readings. As you can imagine, working with a live animal might impose considerable distractions: fidgeting of the animal, the presence of another handler, personal care and self-protection and perhaps an outdoor location with uncontrollable random activity and noise.

Picture reading allows you to start in a controlled environment. Even though you may wish to avoid picture readings in the future, and you may find physical contact to be more enabling and more satisfying, a session with pictures in a quiet location will allow you to go through the procedure and familiarize yourself with the techniques. A few tries at picture readings will help to memorize the step-by-step approaches and will take some of that pressure from you when you finally work with live animals and therefore will help with your confidence level.

Preparation

Collect a few pictures of animals that you do not personally know, preferably of the animal at rest or in a stationary pose. (You don't need to be influenced by its obvious activities.) Write a little information on the back of each of the photos, name of the animal, sex and age only. Pick them all up in a stack and thumb through them quickly selecting only one that captures your immediate attention. Then place the picture in front of you, face down. You don't need to know why that picture captured your attention, in fact it may be better if you just pick one because it gives you an undefined impulse to select it. This will allow you to go through the preliminary preparations without subconsciously presuming and assigning characteristics to the animal before you are ready to start.

Recall the last chapter in which we dealt with "no present time problems". If necessary, reread the chapter to get back to that "in the zone" feeling of freedom from ongoing issues. Relax mentally into a here and now openness.

Now take stock of your body. How does it feel? Get as comfortable as possible. How do your hands feel? Your arms? Your legs? Your stomach? Your neck and shoulders? Got it? This is your base point. Any change from this comfortable state that occurs to you as you look at the animal picture in front of you may result from your impressions from the picture.

Reading the photos

Now turn the picture over and look at it. Note any immediate feelings or impression that you get from the photo. Once you have written down the first impressions, pick up the photo and study it. Record any other thoughts about the animal that come to you, no matter how insignificant they seem to you. Let all of your doubts and considerations go. Do not judge what you pick up as not important or "just your imagination". Use your imagination. Give it free rein, no matter how fanciful it may seem to your normally strict sense of "reality". Go for volume rather than significance. Write it all down. Do not edit or criticize what you have written. Don't even worry about spelling or continuity. Just write it all.

No criticism allowed

The important part of this exercise is not dead-on accuracy, but that you pick up any impressions at all. If you are doing this solo, without any observers or fellow students, do not criticize your written impressions. If you have company in this exercise, fellow students or animal owners, do not accept critiques at this point. Put your observations away for at least a few days. Now, go through the whole procedure again with another picture.

After a week's time, you may wish to review your readings. There may be some surprises: insights that were accurate at the time of the readings or became obvious for the first time after your session. Even if the first exercise appears to be in error, it will serve as a benchmark to show how well you have improved since then. I cannot overstress the necessity of not accepting criticism on your first readings. If you look at your own personal experiences, you must know that you can receive many gracious compliments during the day and only one scathing criticism, and what will you remember from that moment on? The criticism, of course. A critique, even given with "your best interests in mind", can cast doubt on your ability to receive impressions. You do not want that obstacle to discourage you or influence future efforts.

Some assistance
If you have difficulty in getting impressions from the photos, or if you run out of things to write down before you feel satisfied with the reading, ask yourself questions about the animal that can be answered with a "yes" or " no". An either/or impression should be easier to obtain.

Look at the head of the animal and ask:

"Are you happy?

"Are you in pain?"

"Are you in full health?"

Look at various parts of the body and ask:

"Is this part healthy and fully functional?"

"Is this part in pain or unhealthy?"

"Is this problem (if there appears to be one) recent or chronic?" (Either or both)

Now you can get back to more open questions, such as:

"What do you enjoy doing?"

"What makes you happy?"

"What do you love to eat?"

"How do you feel about your owner?"

"Any complaints?"
On these last questions, bear in mind that you are there to give a voice to the animal and that you have an obligation to it to protect it from harm or discomfort. Do not reveal to the owner any information that might aversely affect the relationship unless you feel compelled to believe that the observation will be in the animal's best interests in the long run (such as removing the animal from an abusive or neglectful ownership). Hopefully, you will not be put in such an awkward position until your experience, confidence and animal understanding has developed adequately. When in doubt, you can't go far wrong with silence.

The final instruction in working with pictures is to practice, practice, practice. Go through the full process each time. Be in the here and now (no present time problems). Take stock of your current mental, emotional and physical state. This is your null point. Contact the pictured animal. Changes in mental, emotional or physical condition from your null point may be attributed to the animal. Go for random feelings, messages and visions. Prompt your impressions with either/or (yes/no) questions. Go for volume.

Do not accept critiques or self-criticism.

Good luck and have fun.

CHAPTER SEVEN.
MEDITATION: CONNECTION
THROUGH SPIRIT

The visible world is the invisible
organization of energy.

Physicist Heinz Pagels

Meditation exercises are an important part of this course. Animaltalk can be a very difficult thing to grasp when we are assailed by lifelong training that has convinced our subconscious that it is impossible altogether or impossible for you specifically. To set aside this obstacle, even for a short period, we need an assist that helps to suspend the disbelief and opens up new possibilities. That is the purpose for this meditation specifically timed for this point of the course. I suggest that you use it as often as you wish to reinforce your willingness to accept a new reality.

The same preliminary preparations as mentioned in chapter three apply to this meditation. Prepare your tape or reader as before, and then go through the initial body relaxing preparations.

Meditation Script
Think of your breath. Feel the air moving into your lungs and your exhalation returning to the outside. Think of the air. We are immersed in a sea of air, a huge ocean of moving, flowing atmosphere that extends from the lowest valleys in the earth right out to the stratosphere. The air

is part of the planet. A most vital part. More important to us than even water.

It has mass or weight. At sea level one pound of air takes up 22.4 cubic feet of space. The air above presses down on the air beneath causing a pressure of about 15 pounds per square inch here, just right for us to breathe.

When we breathe this air, it is not a solitary effort. We are sharing the air with every living being on the planet, animals, plants, trees and even the ocean that absorbs air to sustain life under the surface of the seas. When we breathe we are part of the whole eco-system of the planet. We exhale carbon dioxide and moisture that is traded with the plants of the earth. The plants take this air that we exhale and use it as a very essential food that produces growth of the vegetables and fruit that we eat. Our outgoing breath helps trees and all manner of plant life to grow. In this way, we are one with nature and all living things on the planet. We connected to plants that exchange breath with us.

We are connected with animals that share the life-giving oxygen with us.

The air is like a metaphor for spirit. We are immersed in it. We share it with the whole planet, every living thing, every plant or animal and the very soil and rock of the earth itself. Spirit sustains us, defining our lives, just as the air sustains our bodies. In the same way that we share the air, we share spirit with all life on this planet. And that is our connection. What allows us to live, to think and to create also infuses the animals and even the plants on this earth. We are all connected and interdependent.

This is the bridge that allows understanding and communication. As spiritual beings, we know this. It is only the conventional physical limitation of man's egotistical self-definition that denies the unity of life. This is what we hope to overcome. It may be an enormous stretch for you at present to accept this unity through spirit, but that is okay. If it is too much to accept at the moment, we ask for only a small crack in the hard shell of materialistic thinking. If you are not ready to accept it right now, just believe that I can accept it, and let it go at that. That is all that is necessary.

Relax. Do not ponder on what I have said. Just let go and feel connected. Feel how far you can be aware of the space around you. Is it far enough to allow others to share the space with you? Relax your boundaries and without trying to defend against or influence others here in this room let us all be. Peacefully share this room and further if you wish.

<Pause for thirty seconds minimum.>

Slowly, gently return your attention to your breathing. Know that you always share breath, air, and spirit with the rest of the world. Feel your body enriched by your breath and by this experience. Move your fingers and then your limbs feeling the life force in them. Slowly and easily open your eyes, feeling great and wide-awake now. While you are still slightly in the reverie of the meditation, take your pen and record your feelings and observations.

CHAPTER EIGHT.
WORKING DIRECTLY WITH A
LIVE ANIMAL

*Although each of us obviously inhabits a
separate physical body, the laboratory data
from a hundred years of parapsychology
research strongly indicate that there
is no separation in consciousness.*

Russell Targ

Namaste

In many eastern countries "namaste" is a greeting between well meaning individuals. It means roughly, "I recognize the God spirit in you." "I see you as a valid and essential individual who has an equal right to life as I and who has value and purpose just as I have."

Essentially, this is how I greet my animal clients.

Before I talk to animals I clear my mind of personal issues. I recognize the entity that I am going to talk to as an individualized part of the universal life force, soul, spirit, energy. I do not necessarily approach this living being as a horse, a dog, a cat or a snake, but as a perfectly valid essence. Big or small doesn't matter. Male, female or gelding doesn't matter. (That is why I may sometimes mix up the pronouns when I talk to animals, and the owners may quickly correct me, thinking I may be mistaken on other things as well. But it doesn't matter.) Preconceived ideas of levels of intelligence don't matter. I address the essence of the animal, the personalized life energy.

Introduction is important

Before I establish communication, I let the animal know who I am and why I am there. I give assurance that I will not betray a trust by relating to the listening owner any secrets that the animal wishes to keep or anything that might be harmful or might detrimentally effect the its life. Then I ask permission to talk and relate information to the owner. I have had some animal owners object about this, demanding that they have paid to know all that their animal has to say. But I will not betray the trust that I have given the animal, and might refund the money and close off the session immediately rather than do so. More often, because the owner may really have the animal's best interest at heart, I may simply edit the message in such a way as to weed out the disturbing parts. This would be preferable when other parts of the talk would be a benefit to the animal.

Attention and intention

The correct formula for any communication includes attention and intention. Briefly the full cycle of communication involves one entity with a message to be transmitted and another within reach of the first to receive it. They must give each other a degree of attention. The sender must project his message with intention, fully committed to having the message reach the receiver. The communication is completed when the receiver can duplicate the message in exact detail in his own mind. Often a reply is required to complete the communication cycle. Then the original sender should offer an acknowledgement, which will politely terminate the exchange allowing either party to communicate a new thought or sign off on the conversation.

In what passes for conversation today, attention and intention are often overlooked. This results in run-on monologues and feeble attempts to gain attention. I am sure that you have friends who never seem to stop talking and don't even leave room for your response. They are probably suffering from lack of acknowledgement. A firm but friendly "okay", "thank you" or "fine" might bring a smile to their faces and an unusual feeling of completion. Then you might be able to speak in turn.

Animaltalk follows the general rules of communication, and since your animal partner presents an other than normal situation, strict accordance to all of the facets of effective communication will assist you. Give your animal friend your full attention. Don't allow events, other people or critters or even its owner to distract you when you are speaking or listening to it. Give your full attention gently. Do not stare fixedly at the animal (as a predator might), but lower your eyes or look at whatever seems to be attracting the animal's attention. Give it the impression that you are joining with it as a harmless friend.

As for intention, have you ever wondered about the ability of stage actors (before present day voice amplifiers) to have their voices reach the back of the theatre without shouting? Even stage whispers and asides would reach the back rows without interfering with the illusion of secrecy required in the scene. They may have called that "projection" but it depended on intention. The actors "intended" to reach Joe Public, back in the cheap seats, and they did. They focused on the back of the theatre and spoke to the audience there. This attention and intention automatically shifted the vocal cords into a clearer more resonant mode and delivered the words to the target.

Similarly, when sending thoughts to animals, send them with intention. Know that the four-footed client is receiving loud and clear. Project with intention, just as the actors did. If full intention is there, the body and mind will automatically adjust to get the message across.

Find a baseline or null point

Now as I said at the beginning, when you prepare to talk directly to an animal, it is necessary firstly to clear your mind of personal issues. Establish a base line or null point, just as we did previously when we were doing pictures and psychometry. Then anything that you get on contact with the animal relates to the animal. There will be a change point when the animaltalk begins. This may be when you, as the communicator, first touch the animal, or when you mentally tune into that animal essence. At that moment be prepared for any changes in thought, emotions, feelings of illness or wounds, any pictures that may come to mind, any sounds or even changes in the feeling of personal identity. The things that you sense may be in present time or may feel like a memory. Store these first impressions away as things to discuss when the communication is in full swing. Use these feelings or impressions as part of your introduction to the animal. Think, "This is you. Hello, I wish to help you. I am - - " and get into your introduction and promise of confidentiality. Ask permission. Do this as a projection of intention. You intend to be a trustworthy friend, respecting the entity that you are addressing. Project only the desire to help. Offer this opportunity for the animal to possibly improve its life. Be positively persuasive.

Do not humanize

When you are communicating with animals, realize that you are really talking to animals, not odd shaped people. A horse is a horse, a dog is a dog and a cat is a cat. You cannot expect them to do the human thing. Horses are happiest being horses: grazers, herd animals, prey animals with incredible strength and the ability to run (flee). Don't expect them to follow a human viewpoint. They have their own instincts, value system and loyalties. Similarly with dogs: pack animals and occasional predators who happen to have a long history of conditioned living with man. Cats will always be cats: one of nature's best equipped predators with amazing agility and imagination, so efficient that they have lots of time for play and sleeping. Bear in mind the origins of the animals, as a species and as individuals. Stay open to explore their worlds, delightfully new to you. Do not judge them, their thoughts or their actions by human standards or human edicts. Just because an owner thinks that a horse must submit to entering a new trailer, doesn't mean that the horse thinks that it is required of him. This balking horse is not being bad. It is just being a horse, a cautious prey animal that has survived longer than mankind by being suspicious of new environments. The horse may explain to you that this brightly painted dark cave probably has a horse-eating bear inside.

You will find that the animals closely connected to people have amazing understandings and stories about their people. Of course. They are bright and receptive and they hear and see much in their associations with their people. But many of their stories are incomplete, because people act and say strange things beyond a horse's understanding or out of the animal's vicinity. Hence, misunderstandings often resulting in inexplicable behaviour, apparent phobias, spooking, etc.

In such misunderstandings, always give the animal the benefit of the doubt. They have no reason to lie, and in the case of most animals, often do not have the ability to lie. Ongoing problems are often the result of confusion between the animal's basic nature and the human expectations.

Take what comes
Then be open for the response. Throw out any preconceived ideas or expectations. Be a blank slate and accept anything that comes and write it down. Don't tense up thinking that concentration is needed. Relax be open to anything, even if it seems to be "only your imagination". Remember imagination is not a bad thing.

Hopefully you would not get discouraged if your horse doesn't address you in a strong baritone voice like James Earl Jones or tap out messages in Morse code with its hooves. So don't give up if you don't get perfectly clear messages or feelings. Remember that you are attempting to use mental muscles that have probably lain dormant since you were three or four years old. Of course we do not expect that you will suffer from cramps or pulled muscles, but you may get the mental equivalent symptom of doubt or frustration. Just as you would not treat a cramp by straining harder, don't strain against your frustration. Relax. Play with it.

Use imagination

What would be a fun thing to hear from this animal? Does that sound possible? Try again. What if it said " - - " (You fill in the blank.) How does that feel? Write it down, no matter if it sounds ridiculous. Perhaps your automatic real vs. fantasy critic is trying to block what your delicate extrasensory receivers are getting. Try going for the fun stuff first, but give your imagination free rein. Do not look to your past to get your ideas or even judge if it is possible, probable or unlikely. Laugh with it. Anything goes.

Directed questioning

Go back to that first feeling of empathy that you felt for the animal. Did you feel that something was needed? Did you feel strength or weakness? Did you feel happy or sad from the creature? Did you feel energy or sluggishness?

Ask when. Is it now, is it in the past or is it an ongoing (chronic) feeling? Ask why. What caused the feeling, a person, a situation or its environment? Ask who. Who was responsible or just present when this feeling first occurred? Ask where? Is the feeling tied to this present place or somewhere else? Ask how. What caused this or what will make things better?

Write down any response impressions.

Summary of interview techniques

What we have done in the last few comments is try to get around your traditional thinking watchguard. To ask for very general impressions without any preconceived direction may result in apparent failure. If you are among those who have no trouble with that type of unqualified searching, that's great. Go with it. But for those who feel lost without direction, I have first suggested the use of carefree imagination. Anything goes. Nothing will be rejected as unsuitable. Make it light and playful.

Next, if you still have not felt capable of noting a response, let's look at the subject more intensely in a directed fashion, in an analytical format. Hence the reporter's standbys: when, who, where, why and how.

Hopefully one of these tactics will loosen up your reception. There will be no substitute for repetition and practice, so don't let yourself become discouraged. You must stumble and walk before you can run.

Assignment 8-1:

Now, let's go back to looking at animal pictures (or animals if conditions allow) and see what impressions can be picked up. First before you look at the picture or the animal, take stock of how you feel. Take a few easy breaths and search your body for your present comfort level and state of mind. When you know where you are, start your communication.

Remember:

1. See the animal as a spiritual essence, a being as real and viable as yourself.

2. Introduce yourself as a helpful friend who will not betray any confidence.

3. Ask for permission to "talk" in confidence and to relay information to the owner as the animal allows and agrees.

4. Again assure the animal that you are an ally and mean no harm to it.

5. Wait for a response and then question whatever comes up for clarity.

6. Write down your observations. (in actual work with a client, you will be verbally passing on the data as you receive it.)

7. When you finish the interview, thank the animal.

Chapter Nine.

Your Animal Meditation

This story is of a time beyond the memory of man, before the beginning of history, before the beginning of speech almost, when men still eked out their scarce words by gestures, and talked together as the animals do, by the passing of simple thoughts from mind to mind – being themselves indeed still in the brotherhood of the beasts.

H. G. Wells
(from "Stories of the Stone Age" – 1897)

To assist you in experiencing thought exchange with an animal, we will conduct another meditation. It has been said and used in the training of athletes that what has been imagined in detail leaves no less an impact on the subconscious than an actual event. Thus if a training athlete can properly put him or herself into a tranquil state and mentally go through an ideal visualization of their specialty with full feeling and other sensations, the imprint of that potential gold medal performance is etched in their subconscious for future reference. We will attempt to do the same for you with regard to Animaltalk.

We will not pick or suggest a specific animal species for you in this meditation, but will leave the selection to your own subconscious or chance selection. Don't concern yourself about this detail. Just relax when that animal appears and take what comes. No second guessing. Just accept it.

Our meditation will take you through an unusual and more complex trip for a very good reason. We want you to be able to feel that you can move back in time to a period when communication with animals was not so socially spurned.

So let us prepare for another meditation as we have in preceding chapters. Seated comfortably and physically at rest.

Meditation Script

Sit comfortably in your chair and relax totally just as you have done for the previous meditations. Again close your eyes to shut out any visual distraction. Think of your feet, comfortably placed side-by-side on the floor. Feel the slight pressure of the weight of your legs pressing your feet comfortably to the floor. Think of the many small muscles in your feet. They relax, from the tiny tendons in your toes, the muscles supporting the arch and the soles of your feet to the back of your heels. You can feel them move if you wish by spreading and raising your toes, and then letting them relax totally as they rest against the floor. Feel your ankles. Relax. Your calf muscles. Relax. Feel the backs of your knees and your thigh muscles. Relax. Your buttocks and hips relax, almost becoming part of the chair. Your back presses lightly against the back of the chair, and the back muscles relax allowing your spine to rest - properly aligned. Feel your lower stomach relax. You know that your digestion system is functioning effortlessly and comfortably without your conscious consideration. Feel your diaphragm move as your lungs fill and then empty. You know that you are breathing in fresh oxygen-rich healthy air and exhausting carbon dioxide, moisture and air borne wastes. All is well with your breathing and your heart.

Your fingers rest on your thighs or on the arms of the chair. Your wrists are totally relaxed. Your forearms and upper arms rest comfortably. Your shoulders relax and fall lower allowing that bit of tension that you were not aware of to leave your body as the shoulder muscles lengthen and become loose. Let your jaw relax so that it is not clenched tightly closed. There is no pressure between your teeth as your jaws rest and the teeth separate ever so slightly. With the release of tension from your jaws, your whole face relaxes. Those little muscles that control the motion and operation of your eyes relax and you can feel a physical release. Your brows relax and any frowns or lines disappear as your forehead smoothes out.

You have set this time aside for a specific purpose. This is your time. Any sounds that you hear in the background only serve to assure you that the world is operating as it should without your personal attention. And so, when you hear any outside noises, you can relax even deeper, more content in your meditation. Any issues in your life have been put on "hold" for the period of this meditation. Anything that might normally steal some of your attention is "on hold" for this period or is working itself out very efficiently without your concern. This is your well-deserved mini-vacation.

And speaking of vacations, I want you to picture yourself on a special vacation tour of Greece. Greece, the birthplace of civilization, history, mathematics, science and philosophy for the western world. Our contact with the past, and a door to a time before the written history of mankind.

You are with a guided group on one of the Greek islands. You are bored with the guide's memorized account of names, dates and events, and you wish to get a feel of the place for and by yourself. So you wander away from the group. An ancient walled garden, backing up to the shear cliff of the island's big hillside, draws your attention. You enter the garden. Flowers bloom in bright colours everywhere and vines cover the walls and the hillside. You wander deeper into the garden along a clearly defined path of native reddish-colored stone. As you approach the back wall of the steep hillside, you notice a faint trail of white stone partly covered by moss and ground cover. It appears to be seldom used and certainly not very lately.

You decide to follow this old path, and it leads you behind a hedge of thick bushes where it turns directly to the hill. There hidden behind the bushes is an ancient wooden door with heavy cast iron hinges and door handle. You grasp the handle and open the door. It opens onto a spacious but dark tunnel that leads straight into the hill.

You step inside the tunnel and as your eyes become adjusted to the darkness, you notice that the walls glow with a faint luminescence, purple in spots and green in other places. There is enough light to show you the smooth floor and walls reaching off into the distance. Perhaps surprisingly, the tunnel feels comfortable and welcoming, and so you walk into the hill, following the tunnel that curves slowly and ever so slightly downhill. As you travel deeper into the hill you relax even more and feel the promise of a wonderful discovery.

Soon there is daylight ahead and as you approach it you see that the tunnel opens into a small clearing defined by the hill through which you have passed and a semicircle of old trees, a deep ancient forest. This clearing has many of the same flowers as in the other garden and it is open to the blue sky above.

In the middle of the clearing is a stone platform like a centerpiece, a solid flat rock stage. And on that platform waiting for you is an animal!

I don't know what kind of animal it is. It could be a horse, a dog, a cat, a donkey, a mouse or an elephant. But whatever it is, it is magnificent!

Your animal is the very essence of its species - as if it has grown to achieve all of the best qualities of its species and breed. It stands still, majestic in its stance, and you circle the platform to inspect it from all angles. Perfection! The strength and beauty of the animal enthralls you.

It is a mature animal, but not old. It has dignity and pride, but you can feel that it is capable of playfulness and humor. It is fearless and it shows you no reason for you to fear it.

You wish to speak with it, and you remember the instructions from this course.

The first step is easy: finding an admirable characteristic.

You think, "Wow! You are a thing of great beauty!" and send that thought message to it.

Introduce yourself as someone who would be honored to be considered a friend .

Ask its permission to speak with it and assure it that you will respect it and any message that you may receive. You do not impose on it, but you are there if it wishes to share its thoughts. You come as a friend and ally and you are open to receive what it has to offer.

You can feel the energy or the aura of this creature sharing a common space with you and blending with your own identity.

Ask it, "What do I need to know or do to be able to communicate with you and other animals?"

You stand and wait for its advice. In its own timing, it speaks clearly, directly into your thoughts. Wait for it and listen.

<Pause for a full 60 seconds>

Your animal has finished its message and stands observing you as a special friend, an equal. With an openness between your mind and the animal, you thank it for its message and you tell it how much you have enjoyed this experience. <10 second pause> It moves its head in acknowledgement of your remarks, and with grace and a thought message of "Until we meet again," it turns and disappears into the forest.

You turn back to the tunnel in the hillside and enter the cool glowing darkness. As you return through the tunnel, it makes an unexpected turn and you find yourself back here in this room. You feel the weight of your body on the chair, still comfortably relaxed.

In your resting mind, review your experience with your animal. Think and remember:

- What animal did you see?

- Male or female?

- What feelings did you get from sharing a space with the animal?

- What advice did it give you?

- Have you learned anything new about yourself, as seen through the eyes of your animal?

At your own time become fully awake and feeling good. Stretch and open your eyes, and before you say anything to anyone, make some notes to yourself about the experience to allow yourself to recapture the feeling later as you wish.

Using this experience

Now that you have had an experience of connecting with your animal through this meditation, go back and repeat the exercises on communicating with animal pictures and live animals. For a while at least, review your notes on this chapter's experience before each animal contact so that you may recapture the feeling of open mental communication.

If you still feel that you are lacking something that is needed to allow open thought transference, repeat the meditation in this chapter. It does not matter if you contact the same animal, a different breed of the same species or an entirely different creature. Just take what comes and do not judge yourself or the experience. That is the real key. Do not judge. Take what comes and record it or tell it fearlessly, no matter how tempted you are to label it "pure imagination". Practice will take you from this stage to the point of thinking that your reception is plausible and finally to certainty that you are getting a very high degree of accuracy.

It is our fervent hope that students of this book will persevere and become advocates for the animals that they see. We also hope that you will become open to new ways of observing the world about you, and realize that you have powers that you had not dreamed of previously. We wish you good luck and a happier fuller life.

Chapter Ten

Meditation on Success

The greatest discovery of my generation
is that a human being can alter his
life by altering his attitudes.

William James

I find that meditation is a great assistance in maintaining an openness to new experiences. It helps in putting that skeptical self-critic out of the way, at least for the duration of your initial practice sessions. As a bonus for trudging through this course, I offer the following guided meditation that has hidden in its story lots of metaphors and analogies for the road to success in any endeavor.

Arrange your recording or your reader as in the past meditations, and enjoy this trip.

Meditation Script
Relax in your chair, as you did in previous meditations. Call back that feeling of complete calm and restful peace. Check to make sure that you are seated comfortably with your feet apart, arms and shoulders relaxed, your jaw, not clenched, feels no tension and your head rests light and comfortably on your neck. Feel your breath. It is relaxed and natural.

Now you are prepared to take a little vacation. You will leave all of your complications and connections with loved ones and others, behind. Picture this.

You have been staying alone in a little cabin at the foot of a mountain. In the front there is a driveway that leads through lush green meadows from the road up to the cabin that is partially hidden in the woods. The building itself is rustic but comfortable. It blends in with the trees so well you could almost believe that it grew up there with them. Today you are going to complete a dream that you have had since you first came to this cabin. You are going to climb up to the top of the mountain. You have always wanted to do this and so you have prepared and put aside time for yourself to make the full climb today.

It is early morning. A beautiful day. The sun beams down through the trees. A slight breeze promises to keep the temperature just right for the trip. The birds are singing and squirrels scamper about the fringe of the woods. You are making your final preparations. You gather together everything that you could need on this trip: plenty of food, a change of socks, a jacket, a flashlight, fire making necessities, everything that you think that you could possibly use. Next you select a backpack that can hold all this stuff and you pack it neatly. That's it. Everything that you could need is in the pack, and you buckle it closed. Next the pack goes on your back and you pick up a walking stick and head out the door.

You lock the door and turning about, you stride up the wide path into the woods. The air is cool among the trees and the sunlight filters down through the green treetops to the trail. Birds chirp and sing, appearing and disappearing in erratic darting flights across the pathway.

In the dry leaves at the foot of the trees small animals, probably squirrels, mice or rabbits, scamper about on their daily business, making noises louder than you would expect, but the sounds are soothing. These little creatures are not disturbed by your presence; they do not cower in fear of you. It is as if they were accepting you into their world.

You continue up the hill and the path progressively become steeper and narrower. There are places where the path cuts back on itself to gain the height of a new cliff and occasionally there are times when you must use your walking stick for that extra push, or you scramble a few meters on all fours. The ground becomes rockier and at times you find that the path leads through large cracks in solid rock. The backpack seems to gain weight and the straps dig into your shoulders. Your legs are starting to burn, your back feels beat up and your breath has become a steady panting.

You look at your watch it is still slightly before noon. At the next clearing you look down the hill. You have traveled a long way, but as you glance up to the heights, you realize that you have not yet reached the halfway point. As you plod upward you decide that now is the time to take a break. Soon you reach a clearing that opens on a view of the creek that tumbles down the mountainside. Actually there is a waterfall of about fifteen feet here where the narrow band of water slips over a rocky ridge and cascades down to a dark pool producing a misty spray and a white noise that blankets out any disturbing sounds. The cool dampness has produced a healthy moss that covers fallen wood and rocks in your clearing and there is a good-sized level shelf of stone overlooking the picturesque falls. The morning sun filtering down through the fringe of trees surrounding this open area has warmed the rock.

You struggle out of your pack and sit on the soft moss with your back against a moss-covered stump. You reach into your bag and find a small blanket that you spread out before you. Digging deeply into your bag you select a delicious lunch and a refreshing beverage. You eat slowly feeling your muscles relax as you enjoy the tranquility of the place. When you are finished eating you stretch out on the mossy stone and surrender your body to the warm peacefulness of this beautiful picnic spot. You doze for about twenty minutes in total relaxation.

<Pause for about twenty seconds.>

You slowly stretch the final stresses from your body and quietly get up and walk over to the cascading creek. Taking an enameled cup, you reach out to the falling water and catch some of the cool clear water. It is totally refreshing and rejuvenating as you savour the purity of this fresh water. You look to the top of the mountain and recapture your desire to reach the top today.

You go back to your backpack and empty it out onto the ground. With great consideration, you start to separate all of the items into two piles: things that you know you will need (like a bottle of water, a light jacket, and so on) and things that you don't need. You discover that you can leave most of the stuff, and most of the weight, behind. You pack the bare necessities into a very light load and sling it onto your back. You know that all that you have left behind will not clutter this beautiful spot. Animals will consume any edibles very quickly. Other travelers will appreciate much of your discards, and the next rainfall will wash the remaining stuff down the creek, shredding it on waterfalls and rocks and absorbing it into the soil as food for the vegetation.

You turn back up the hill on the path and soon you are deep along the wooded trail, progressing swiftly, with a spring in your step and a carefree floating feeling, thanks to your brief timeout and refreshing lunch.

Soon you come to the end of the treeline. The sun beams down on the mountainside and a cool breeze conditions the air, refueling you with new energy. There are very few birds here. No tree squirrels or rabbits. Only little ground squirrels that stand in curiosity to see you stride past. You are surprised when the clear blue air suddenly becomes misty, and you seem to walk into a deepening fog. You have to bend over slightly to examine and follow the faint outline of your path. It seems that very few people have climbed this far and the path has petered out to an old trail that would have long vanished if the moss and ground cover were not so delicate. You persevere and soon the air clears and there before you only one hundred yards away is the summit of the mountain. You have just passed through a low hanging cloud that the breeze has brushed against the hillside as it searches for a clear path through the sky.

You stride to the very peak of the mountain and standing on its bare rock surface, you turn and look around. You can see for miles in all directions. 360 degrees of clear unobstructed scenery. You can look back in the direction that you have come and see a small cloud break away from the hillside and float downwind as a puffy white shape-shifting cottonball. The stream and the waterfall are down in that direction and your cabin is "way down there".

You sit in the sunlight and absorb as much of the beauty as you can. No words are necessary. The beauty has to be seen to be appreciated. You have fulfilled your quest, and it is all worthwhile.

Soon you are aware that the sun has dropped in the west and is about to disappear behind the next range of mountains. It is time to head home. You are not concerned, because it is all downhill and you know the way very well.

One last glance around at the fantastic view and then off you go, retracing your steps down through the alpine meadow, down through the woodlands, past your noontime stopover, and on to the base of the mountain and your cabin.

In a moment, you have a roaring fire in the fireplace, you have gone to the refrigerator for a cold drink and a snack and then you sit at your desk to record your day's events in your journal. The words flow from your pen and in no time at all you have filled more than three pages with details of your adventure. You lean back and relax in satisfied pleasure.

Tomorrow will be another day. You will see the same people that you saw last week. You will do much of the same things that you did last week. But it will be different, because you have completed a quest that you had set for yourself. You are the same but you are renewed. And that is great!

Carrying that feeling of successful completion, your consciousness returns to this room, here and now. You feel great: relaxed, healthy and capable. You can recapture that feeling of contentment and success any time that you wish, by simply closing your eyes for a moment and recalling that climb.

LAUREN BODE was raised in the back country of South America where she thrived in an environment filled with animals, but free from the mental and social limitations that might have been imposed on her unusual abilities in a more "civilized" existence. Horses, cattle, domestic pets and more exotic creatures from the surrounding jungle were part of Lauren's daily life in her early years, and her love and empathy for animals has remained constant to this day.

Lauren "talks" telepathically with all types of animals: horses, dogs and cats, of course, but also mice, lizards, snakes, birds, etc. Lauren runs very successful workshops on animal communication for animal lovers who can appreciate the unfathomed abilities and intelligence of their non-human friends.

She contributes regularly to Canadian horse magazines and has been featured on local and national television. The news media has called her to pick the winners of featured Ontario horse races where she never disappoints them and often converts skeptical commentators into "believers".

Lauren does a lot of travelling through Canada, the United States and overseas for direct personal contact with her animal subjects, but much of her animal communication is done over the telephone or on computer. She works from photos mailed or emailed to her in advance of the readings. For more information check her website at www.animaltalk.ca.

JOHN BELL likes to consider himself a fledging mystic, but he cautions readers about the meaning of the word. To him, a mystic is a person who believes and acts on the assumption that what happens in the physical world has its origins elsewhere, in spirit and thought.

After forty-five years of age, while employed as a "full time" manager, John managed to obtain a Ph.D. in philosophy, complete three years of religious study, become a co-founder of a church in B.C., lecture in Canada and the U.S. on spiritual and psychic development, contribute articles to magazines and author a how-to book on psychic ability. He also took part in studies on precognition (in Alberta) and telekinesis (York University, Ontario).

During his business career, his forte was starting up and successfully running branch sales offices for Ontario manufacturers. Despite cautions from associates, he made no attempt to hide his spiritual studies and beliefs from his employers and clients. He says everyone remembers a "kook" and that is fine for business.

Now semi-retired, John assists his wife, Lauren, in her career as an outstanding animal communicator.

Printed in the United States
114670LV00001B/7-27/A